Earth Explorations

The Water Cycle

Jenny Karpelenia

PERFECTION LEARNING®

Editorial Director:	Susan C. Thies
Editor:	Mary L. Bush
Design Director:	Randy Messer
Book Design:	Tobi Cunningham, Brianne Osborn
Cover Design:	Michael A. Aspengren

A special thanks to the following for their scientific review of the book:
Kristin Mandsager; Instructor of Physics and Astronomy;
North Iowa Area Community College
Jeffrey Bush; Field Engineer; Vessco, Inc.

Image Credits:
©Dave G. Houser/CORBIS: p. 9; ©Martin Harvey; Gallo Images/CORBIS: p. 13;
©Annie Griffiths Belt/CORBIS: p. 26; ©Peter Johnson/CORBIS: p. 27; ©Cheque/CORBIS: p. 31;
©Jim Reed/CORBIS: p. 33; ©Terry W. Eggers/CORBIS p. 41

Photos.com: front cover, back cover, Try This! backgrounds, pp. 3, 5, 6, 7, 8, 10, 11, 12, 14, 15, 16, 17, 19, 20, 21, 22 (top), 23, 24, 25, 29 (center and bottom), 32, 34, 35, 37, 38, 39, 43, 46, 47, 48; Ingram Publishing: p. 22 (bottom); PLC images: pp. 4, 18, 28, 30, 36, 40, 42; Digital Stock: pp. 29 (top), 45

Text © 2005 by Perfection Learning® Corporation.
All rights reserved. No part of this book may be reproduced, stored in a
retrieval system, or transmitted in any form or by any means, electronic, mechanical,
photocopying, recording, or otherwise, without prior permission of the publisher.
Printed in the United States of America.

For information, contact
Perfection Learning® Corporation
1000 North Second Avenue, P.O. Box 500
Logan, Iowa 51546-0500.
Phone: 1-800-831-4190
Fax: 1-800-543-2745
perfectionlearning.com

1 2 3 4 5 6 PP 09 08 07 06 05 04
ISBN 0-7891-6324-1

Table of Contents

1. Splish Splash 4
2. A Few Words About Water 7
3. Round and Round 14
4. Going Up 19
5. Cooling Off 24
6. Falling Down 30
7. On the Ground 34
8. The Cleaning Cycle 38
   Internet Connections and Related Reading
   for the Water Cycle 44
   Glossary 45
   Index 48

Splish Splash

Summer vacation has finally arrived! No more school. No more teachers. No more homework. It's just you and your friends and endless days of fun.

You hop out of bed and make a bathroom pit stop. You shower and then brush your teeth in the steamy bathroom. After getting dressed, you head downstairs.

Mom is boiling water for poached eggs. She hands you a glass of juice and asks you about your plans for the day. She reminds you to finish your chores before anything else.

After breakfast, you gather your dirty clothes for the laundry and load the dishwasher. You fill the cat's food and water bowls. Then you head out. You walk a few doors down to see your best friend. The two of you decide to shoot some hoops. After a half hour, you're hot and sweaty, so you decide to cool off with the hose. That reminds you that you were supposed to water the flowers at home.

The two of you head back to your house to water the flowers and pick up your bike. After spraying the yard—and yourselves—with water from the garden hose, you hop on your bikes. Picking up another friend, you ride along a river path. At one point, you stop and skip stones across the water.

When you get hungry, you ride home. You grab three bottles of water out of the fridge and hand one to each of your friends. While munching on sandwiches, the three of you decide to spend the hot afternoon at the pool. On the way, you swerve your bikes through every yard sprinkler.

At the pool, you and your friends go down the giant water slide. You take turns diving and doing cannonballs off the board. About midafternoon, you notice dark **clouds** moving in. Soon lightning streaks across the sky, and the lifeguards clear the pool.

On the quick ride home, you get soaked by a sudden downpour of rain. Mud splashes your legs from the puddles forming on the ground. Small rivers of water rush down the street gutters.

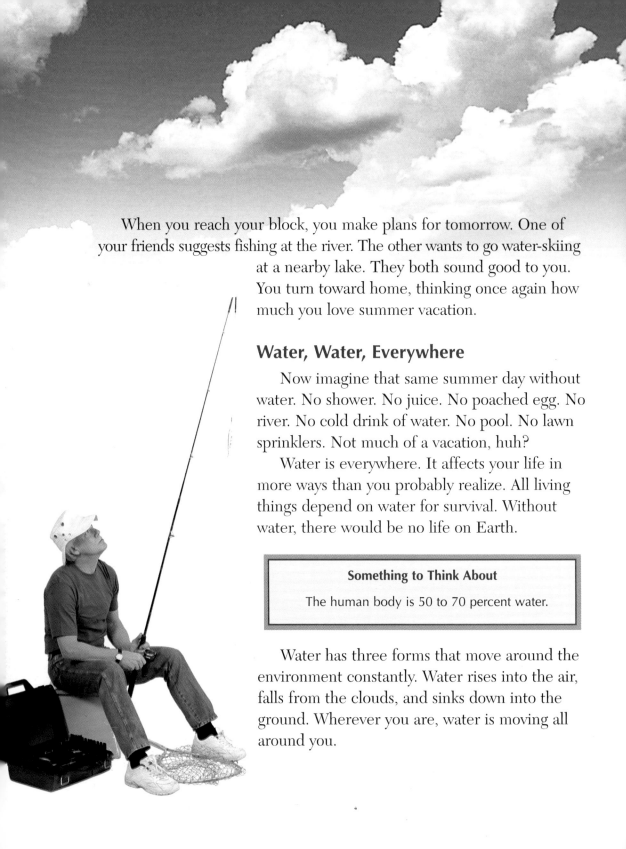

When you reach your block, you make plans for tomorrow. One of your friends suggests fishing at the river. The other wants to go water-skiing at a nearby lake. They both sound good to you. You turn toward home, thinking once again how much you love summer vacation.

Water, Water, Everywhere

Now imagine that same summer day without water. No shower. No juice. No poached egg. No river. No cold drink of water. No pool. No lawn sprinklers. Not much of a vacation, huh?

Water is everywhere. It affects your life in more ways than you probably realize. All living things depend on water for survival. Without water, there would be no life on Earth.

> **Something to Think About**
> The human body is 50 to 70 percent water.

Water has three forms that move around the environment constantly. Water rises into the air, falls from the clouds, and sinks down into the ground. Wherever you are, water is moving all around you.

A Few Words About Water

A World Without Water

About 4.6 billion years ago, Earth formed from the accumulation of particles in space. After millions of years, the mass of particles became a planet. But the planet was a steaming, violent place without oxygen. Temperatures were too hot for water to exist in a liquid form. **Water vapor** floated in the air but couldn't cool off enough to become a liquid.

Eventually, the Earth began cooling off. Water vapor changed to liquid water. Rain filled the lower spots on the planet's surface. It seeped into the cracks in rocks and flowed underground. This was the beginning of the **water cycle** on Earth.

How Much Water?

Water covers about 70 percent of the Earth's surface. For this reason, Earth is often called "the blue planet." From outer space, the planet looks blue since the majority of it is covered with water.

Approximately 326 million cubic miles of water flow on Earth. A cubic mile is a cube with sides that are a mile long. More than 1 trillion gallons of water can fit into one of those cubes. The Earth has 326 million of those cubes filled with water.

Even though there is so much water on Earth, very little of it is actually good for human use. About 97 percent of the water is **salt water**, which humans cannot directly consume. The other 3 percent is **freshwater**. However, of that 3 percent, only 1 percent is available as running water. The other 2 percent is frozen in glaciers and **ice caps**. So out of all the water on the planet, only 1 percent is naturally good for human use.

Imagine This!

You've just mowed the lawn on a hot, sunny day. You're dying of thirst. Inside your house, 100 glasses of water sit on a table. Of those 100 glasses, 97 of them contain salt water that you can't drink. Two of the glasses are filled with water that's frozen solid. Only one glass has fresh liquid water that you can drink. You guzzle that glass down and stare at the other 99 useless glasses. Now you have an idea of what the world's water supply looks like.

Salt Water

The majority of the water on Earth is salty. The oceans, seas, and even a few lakes are filled with salt water. Where does the salt come from? Salt is a mineral found in rocks. Over time, rain and rivers wear down rocks. The salt is released and dissolves in the water. Eventually, this water flows into the oceans. Ocean water is about 3.5 percent salt.

The Great Salt Lake

The Great Salt Lake in Utah is one of the few saltwater lakes. This lake has about 15 to 25 percent salt in its waters, which makes it much saltier than the ocean. Water enters the lake from four rivers and several streams. There are no rivers leading out of the lake, however, so all of the water is trapped. When water from the lake evaporates (changes to a gas), it leaves the salt behind. The trapped water keeps getting saltier and has nowhere to go.

Salt water is not safe for humans and most animals to drink. When a body takes in salt water, its tiny **cells** react. Cells are made mostly of water with a small amount of salt mixed in. Tears and sweat are made of this slightly salty water. Taking in large quantities of salt water, however, is harmful. When this does happen, the water inside the body's cells tries to balance things out. It rushes out to **dilute** all the salt water. This causes the cells to shrink, shrivel, and dry up. The body becomes dehydrated and can eventually die.

It is easier to float in salt water because the water is denser than freshwater.

Salt water can be made safe for use through a process called *desalination*. Desalination removes the salt from water. It is used in areas near oceans or large seas and in countries that have very limited freshwater supplies. However, currently desalination uses tremendous amounts of nonrenewable, expensive energy sources, so it is not a good solution for everyone. Finding better sources of renewable energy may make desalination more common in the future.

Freshwater

Freshwater is water that contains very little or no salt. Freshwater fills most lakes, rivers, streams, and ponds. It also flows underground and on the surface of the Earth. Freshwater sources are important to much of the world's population. They depend on this water for drinking, washing, running industries, and watering crops.

Much of the Earth's freshwater is frozen in glaciers and ice caps. Glaciers cover about 10 percent of the Earth's surface. While these huge sheets of ice contain millions of gallons of freshwater, this water can't be used by humans, animals, or plants.

TRY THIS!

Build your own model of the Earth's water supply.

Measure 100 ml of water in a metric measuring cup or graduated cylinder. Pour 2 ml of that water into a baby food jar or small glass. Pour another 1 ml of the original water into another jar or glass. Pour the remaining 97 ml into a large jar or glass container.

Add a tablespoon of salt to the large jar and stir until dissolved. You can also add blue food coloring if you'd like. Freeze the jar or glass containing 2 ml of water. When all the jars are done, ask your classmates, friends, or family to guess what each jar represents. Then label the jars correctly.

Conservation and Pollution Solutions

Since so little of the water on Earth is fit for use, it is important not to waste it or pollute it. Conserving water means using only what you need. Take short showers and make sure the dishwasher is full before running it. Don't let faucets run or even drip. A slow steady drip over hours or days can send gallons of water down the drain. Water your yard in the morning when a smaller amount of water will be more

effective. (Heat causes water to evaporate into the air instead of soaking into the soil.) Taking steps to conserve water will protect the future of freshwater.

The limited supply of freshwater on Earth is also threatened by pollution. Oil spills, litter, soil, soaps, pesticides, fertilizers, and household and industrial chemicals contaminate freshwater supplies. The correct use and disposal of harmful materials can protect precious freshwater sources.

This penquin was caught in an oil spill. These accidents can be deadly to birds. The oil destroys their water-repellent feathers, leaving the birds defenseless. Sometimes they also swallow the oil, which poisons them.

3

Round and Round

Eighty million years ago, a flying reptile called the *pteranodon* drank from a watering hole. You might be drinking some of the same water **molecules** today. Fifteen thousand years ago, a woolly mammoth stood on a layer of glacial ice. Some of the water molecules trapped in that glacial ice could be in the next raindrops that fall in your town. Almost 230 years ago, General George Washington led his troops across the cold Delaware River to battle in the American Revolution. Some of the water molecules that were rushing through the Delaware River then may be running through your faucets right now. These "recycled" water molecules may be used to wash your clothes, your dishes, or your hands.

How is this possible? Through the water cycle. The water cycle, or hydrologic cycle, is the constant change and motion of water molecules on the Earth's surface and in the air. Water moves from the oceans to the air to the ground and back again in a continuous cycle. Sometimes the process takes only a few days. Other times, the process takes hundreds or thousands of years.

Millions of Molecules

Water is made up of small molecules, or particles. Alone they are invisible. But bunched together in millions, billions, or trillions, water becomes visible to the human eye. One molecule of water is made up of smaller parts called *atoms*. Two hydrogen atoms are stuck to one oxygen atom. That's why the formula for water is H_2O.

Solids, Liquids, and Gases

The tiny molecules that make up water are constantly changing from one form, or state, to another. The atoms and amounts do not change. But the position of the molecules and the speed that they move change. It is this ability to change that allows water to recycle from land to air to sea.

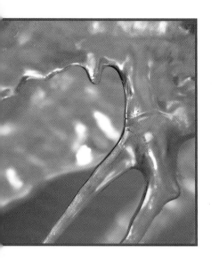  

Water has three states—solid, liquid, and gas. The position of the molecules is different in each form. In solid form, the water molecules are arranged in a particular pattern. In liquid form, the molecules are released from their solid position and can move around. In gas form, the molecules are scattered far apart.

The molecules in solid water (ice, snow, frost) are in a fixed position. They may shake slightly, but they don't move. Liquid water molecules flow freely, sliding past one another. They are constantly changing positions. Molecules of gaseous water (vapor, steam) fly and scatter and bounce off one another. They move quickly to spread out in any given space.

What causes the molecules in water to change from one form to another? Temperature. When heat is added to a solid, it causes the molecules to speed up and move apart. Glacial ice becomes water this way. When more heat is added, the molecules move farther apart at even higher speeds and become a gas. Ocean water changes to water vapor in this manner. Loss of heat causes opposite changes. When gas molecules cool off, they slow down and move closer together, forming a liquid. Vapor cools into droplets of water. At colder temperatures, the molecules in those droplets slow down even more and "freeze" in a fixed position. The droplets now become icicles.

Skipping a State

Sometimes a state of matter can be skipped. For example, a solid can skip the liquid form and go right to a gas state. This is called *sublimation*. Some glacier ice can change from a solid to water vapor without a liquid phase in between.

A gas can also change directly to a solid. This is called *deposition*. Water vapor can touch a very cold window and turn right into ice crystals.

The Water Cycle

As water molecules change from one form to the next, they move through the water cycle. Liquid water in the ocean changes to vapor (gas) in the air, which changes back to rain or snow in the clouds. This **precipitation** flows across the ground as water until it returns to the ocean to begin the cycle again. This endless cycle is occurring every minute of every day, while the total amount of water on Earth remains the same.

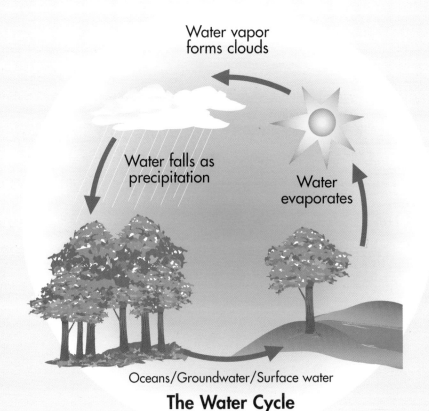

The Water Cycle

Going Up

Evaporation

The Sun heats the water in oceans, lakes, rivers, and even puddles. The molecules near the surface of the water move quickly. They have so much energy that some escape from the liquid and become water vapor in the air. This change from a liquid to a gas is called **evaporation**.

The evaporation process is similar to popping popcorn. The heat from the microwave or stove makes the water inside each little corn kernel heat up. It boils and turns to steam. The pressure blasts open the kernel and cooks the tasty white part of the corn inside. If you were to open the bag or remove the lid while the corn was still cooking, the kernels would go flying all over as the water molecules burst apart. The kernels closer to the top would escape first. This is similar to the evaporation of water. Evaporating water molecules near the surface escape from the bonds of the other water molecules. They make a break for it and float freely in the atmosphere as water vapor.

19

Steam Versus Water Vapor

When you take the cover off a pot of boiling water, steam escapes. Steam is a gaseous form of water. Is this the same thing as water vapor? Not exactly. Steam is gaseous water that forms as a result of boiling, or reaching a temperature of 212°F at sea level. Water vapor is gaseous water that forms at lower temperatures.

Most evaporation in the world takes place in the oceans. After all, oceans cover most of the planet. Since the oceans are salty, is the water vapor coming from the oceans salty as well? No, the evaporation process leaves the salt behind in the oceans. Over time, this makes the oceans saltier as more salt is dropped off by rivers and left behind during evaporation.

TRY THIS!

See for yourself how salt is left behind during evaporation. Pour a small amount of water into a cup. Add some salt to the water and stir. Pour some of the salt water onto a plate so you have a thin puddle. Set the plate in a warm spot—on the sidewalk, under a sunny window, or under a lamp. Check the water occasionally. What happens to the water? What's left behind on the plate?

After a rain, sidewalks and driveways are wet. Puddles may form. Why don't these areas stay wet forever? Because the water molecules heat up, move faster, and become water vapor. Evaporation dries the wet areas as the water moves into the air.

TRY THIS!

On a warm day, dump a glass of water on a sunny sidewalk. Use a piece of chalk to outline the puddle. Check the puddle every once in a while for several hours. What happens to the water inside the chalk outline?

You should notice that the puddle will shrink inside the outline until it finally disappears. That's because the water has evaporated into the warm air.

Ever hung up wet clothes to dry? Thanks to evaporation, you can have dry clothes in a matter of minutes or hours, depending on the temperature and the amount of water vapor already in the air. Warm air hitting the wet clothes makes the water molecules move faster. This causes some of the molecules to break free and become water vapor. After a while, all of the water will have turned to vapor and the clothes will be dry. Clothes dryers work the same way. The warm air circulated by the dryer causes the water to evaporate from the wet clothes. The water vapor is vented outside and joins the rest of the vapor in the air.

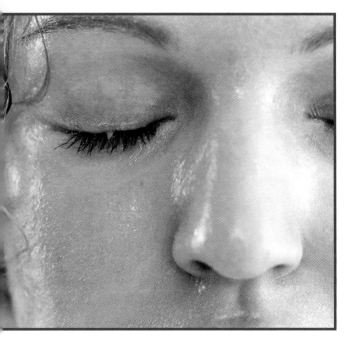

Evaporation takes place in humans (and many other animals) too. When you get out of the shower, the water on your skin evaporates in the warmer air. When water evaporates, it takes heat with it. This is why you shiver or get goose bumps as the water evaporates from your skin.

Sweating creates the same effect. When sweat evaporates off your skin, it takes heat with it. The water vapor "disappears" into the air, leaving you cooler.

Transpiration

All of the water vapor in the air doesn't come from bodies of water or animals. Some of it comes from plants. The evaporation of water from plants is known as **transpiration**. Transpiration takes place on the leaves of plants. Thousands of tiny openings called *stomata* are found on the undersides of leaves. These stomata are little doorways that let carbon dioxide gas in and water out. Plants pull up water out of the ground through their stems and use this water, along with carbon dioxide, to make food. Excess water escapes out of the plants through the stomata. This water evaporates into the surrounding air.

Some areas of the world, such as rain forests, have huge numbers of trees and plants. Here transpiration puts large amounts of water vapor into the air, creating a **humid** environment. As the vapor moves through the water cycle, this high humidity leads to steady rainfall.

Evapotranspiration

Sometimes the processes of evaporation and transpiration are linked together. Evapotranspiration is the combination of both processes that change liquid water to water vapor.

5

Cooling Off

When water vapor cools, it turns into a liquid. This process of changing from a gas to a liquid is known as **condensation**. If you've ever had an ice cold drink on a hot day, you've seen condensation occur. Let's say you pour a glass of frosty lemonade. A short time later, your glass is wet on the outside. Where did that water come from? It's condensation. Water vapor in the air touched the cold glass. The water vapor cooled down and turned into water droplets.

Cloudy Condensation

Condensation often forms clouds. Clouds are visible bunches of tiny water droplets. You see these clouds on the ground and in the sky.

In a Fog

Fog is also visible bunches of tiny water droplets. The only difference between the two is how far above the ground they are. Generally, in scientific terms, condensation 50 feet or less above the ground is called *fog*, while condensation higher than 50 feet is called *clouds*.

On the Ground

The little cloud coming out of your house when the dryer is running. The foggy moisture that covers your bathroom mirror after a shower. The puffs of breath you blow when it's cold outside. The fog that covers your glasses so you can't see. What do all of these things have in common? They are all forms of condensation that occur near the ground.

When a clothes dryer evaporates water from wet laundry, the water vapor is vented outside through a tube. When the vapor hits cool air, it condenses and forms a cloud.

After you take a hot shower, the mirror in the bathroom is usually foggy. That's because some of the warm water from your shower changed into water vapor that filled the bathroom. When the vapor came in contact with the cooler mirror, it condensed, forming a layer of tiny droplets. In time, the fog will evaporate from the mirror. (Unless you use your hair dryer to speed up the process so you can see yourself in the mirror!)

When you exhale (breathe out), water vapor leaves your body. Warm, moist air from your lungs gets pushed out with every breath. On a cold day, this warm air condenses quickly and forms a little cloud in front of you.

Eyeglasses are a perfect target for condensation. They "fog up" when the wearer enters a warm building on a cold day, opens a dishwasher that's just finished running, or checks on dinner cooking in a warm oven. All of these actions lead to condensation. The warm air in the building contains water vapor. Some of those vapor molecules stick to the cold glasses and condense. The steam from the dishwasher and hot air from the oven condense when they meet the cooler glasses. Luckily, the fog usually evaporates quickly, allowing for clearer vision.

If temperatures are cold enough, water vapor can condense on beards and then freeze.

In the Sky

Condensation also forms clouds in the sky. Clouds are actually billions of tiny droplets of liquid water. Warm air containing water vapor is lifted upward by **convection currents**. As that warmer air climbs higher, the temperature drops. This cooler temperature causes the water vapor molecules to slow down and change back to a liquid. The little droplets of water stick to dust and smoke floating in the atmosphere. These particles are called *condensation nuclei*. They give the droplets a center to hang on to. As more and more water droplets stick to the particles, a cloud is formed.

Convection Currents

The molecules in warm air move faster and are more spread out than the molecules in cooler air. This means that warm air is less dense (less tightly packed together) than cool air. Because of this, the "lighter" warm air moves upward above the "heavier" cool air. When the warm air cools, it moves downward again. This cycle of warm and cool air movement is called *convection*. Convection currents are the flow of warm and cold air in the atmosphere.

TRY THIS!

Make your own cloud in a bottle.

Materials
water
clean 2-liter pop bottle
matches
clay
bicycle pump

Directions

1. Put a small amount of warm water in the bottom of the bottle and swirl it around. Pour the water into a sink. This will leave water vapor in the bottle.

2. Ask an adult to light a match. After a few seconds, blow it out and drop it into the bottle. The smoke from the match will create condensation nuclei.

3. Quickly put the clay on top of the bottle opening to trap the water vapor and smoke inside. Then push the stem of the bike pump through the clay and into the bottle. Make sure the clay stays sealed around the opening.

4. Pump air into the bottle. You should feel the pressure building. Increasing the pressure inside the bottle has the same effect as cooling down the temperature.

5. Look inside the bottle. What's happening? The water vapor should be condensing and collecting on the smoke particles.

6. Take off the clay plug and watch a cloud form in the jug.

7. If desired, repeat the steps to make more clouds.

Three basic types of clouds form in the skies—cirrus, cumulus, and stratus. Cirrus clouds are thin, wispy clouds of ice crystals found at high **altitudes**. Cumulus clouds are fluffy clouds that look like cotton. If these clouds grow bigger and darker, they become cumulonimbus clouds. These clouds produce thunderstorms and tornadoes. Stratus clouds are layered clouds that blanket the sky. Near the ground, these clouds are often called *fog* or *mist*.

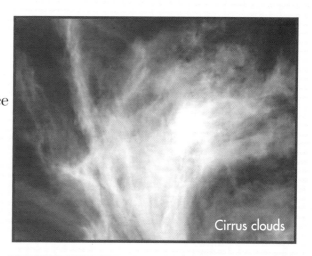

Cirrus clouds

Cumulus clouds

Stratus clouds

Trails of Clouds

Jet airplanes often leave trails of clouds behind them. These trails are called *condensation trails* or *contrails*. Hot gases and water vapor are pushed out the back of a jet engine, making the plane move forward. The hot gases heat the water vapor in the air behind the jet. The water vapor condenses into liquid water droplets, which form cloud trails behind the jet.

6

Falling Down

Falling from the Skies

When clouds begin to form, they are light and airy. Air currents keep them floating in the sky, and wind moves them around. Eventually, however, the mass of droplets gets too large and too heavy. The air currents keeping them aloft are no match for their weight. Gravity wins the battle, and the droplets head toward the ground. These falling water droplets are called *precipitation*. Depending on the temperature, precipitation can take several forms—snow, sleet, freezing rain, rain, or hail.

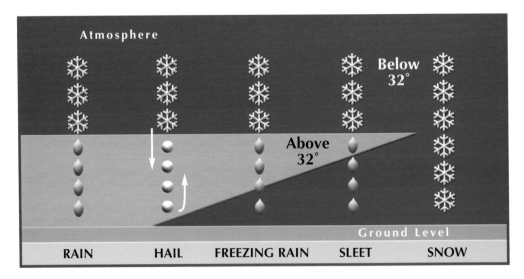

TRY THIS!

Conduct your own condensation/precipitation experiment. Pour about an inch of hot water into a glass jar. Turn the lid of the jar upside down and set it on top of the jar. Fill the lid with ice cubes. Watch the underside of the lid for 10–15 minutes. What happens?

You should see condensation forming on the lid. Water molecules in the warm water will evaporate and rise in the jar. When this water vapor hits the cold lid, the molecules will condense and change back into a liquid. After a while, the drops will get too big and too heavy and will fall back into the water. This is like raindrops falling from the sky.

Frozen Solid

At cold temperatures, the water vapor in the clouds condenses directly into ice. This ice forms crystal structures on dust or smoke particles. The snow crystals continue to grow as they collect more water molecules. Eventually, they become too heavy and gravity sends them downward. If the temperatures near the ground are also cold, the precipitation will remain frozen, and the water will return to Earth as snowflakes.

If the temperature hovers around the freezing point (32°F) near the surface of the Earth, sleet can occur. Sleet is rain or melting snow that freezes or partially freezes as it nears Earth. Sometimes sleet is a combination of rain and ice, and other times it's all ice.

Freezing rain is different than sleet. First the snow falls through warmer air and melts. Then the rain passes through a colder layer that supercools the drops. They remain in the form of a liquid but freeze as soon as they hit a cold object on the Earth's surface. Trees, power lines, houses, roads, and sidewalks are good targets for freezing rain.

Raindrops Keep Falling on My Head

If the temperature near the ground is above freezing, rain will fall. Raindrops vary in size. As they fall through the atmosphere, they can smash together in the wind currents. Some drops join together to create one larger drop. Others break apart from the force of the blow and become several smaller drops.

It's Raining Hamburger Buns

When most people draw a raindrop, it looks like a triangle with a rounded bottom. In reality, raindrops start out as perfectly round spheres and end up looking like hamburger buns. As they fall through the sky, the round drops push against the air. This makes the bottoms flatten out, resulting in a bun shape.

TRY THIS!

Try this simple experiment to see how the size of raindrops varies. Cover a cookie sheet or tray with a thin layer of flour. Hold the tray in the rain for a few seconds—just long enough for a few drops to land on the sheet. (Too much rain will make a doughy mess.) Bring the sheet inside and measure the size of the "holes" left by the raindrops. What do you notice?

Great Balls of Ice

Hail makes an interesting journey from the clouds to the ground. Hail occurs when ice balls fall from a cold cloud and hit warmer air below. The warm air is then carried upward by currents, taking the ice balls with it. More moisture freezes to the balls, making them larger in size. This process can happen many times. Each time the balls return to the clouds, they become larger. Eventually the ice balls are too heavy to be carried upward by the warm air, and they crash to the ground as hailstones.

Hailstones are usually about the size of peas, but they can be as large as grapefruit. These ice pellets have rings similar to a tree trunk. Each ring represents another layer of ice picked up in a cloud.

7

On the Ground

Whether it's in the form of rain, snow, or ice, all precipitation eventually reaches the ground. Much of this water lands in the ocean, which covers a large area of the planet. But what about the rest of the freshwater? It ends up as **groundwater** or **surface water**.

Groundwater

When precipitation hits the ground, much of it is absorbed into the soil. Just below the surface, plant roots soak up needed freshwater. Water not used by plants continues to move deeper underground. It filters its way through and around rocks and particles of soil. Many rocks have spaces in them that the water can seep into. Rocks such as sandstone and limestone allow water to flow through them.

Eventually the water will hit a layer of rock that doesn't allow it to pass through. Clay and shale are such rocks. These rocks act as a barrier to the flow of water.

> ### Making Groundwater or Coffee
>
> The downward movement of water through rock and soil is called *percolation*. Percolation is also how a coffeemaker works. Water is poured into the top of the coffeemaker and percolates down through the ground coffee beans and into the pot.

These layers of rock establish **aquifers**. Aquifers are areas of rock and soil that hold groundwater. They rest above layers of rock that stop the flow of water. The top of an aquifer is called the *water table*. The level of the water table changes as more or less water is absorbed into the ground. When an aquifer becomes saturated (can't hold any more water), the water table will rise with the addition of new water. If water is scarce, the water table lowers.

TRY THIS!

Build your own groundwater model. Use two sponges of the same size. Place one on top of the other. Pour water on the top sponge until it's saturated and water begins to flow into the second sponge. This shows how water travels down through spaces in rock.

Now squeeze out the bottom sponge and place it in a baggie. Seal the baggie. Put the other sponge back on top and run water over it again until it's saturated. Where does the extra water go now?

You should notice the excess water seeping out the sides of the sponge. The sealed sponge is your rock barrier. The top sponge is your aquifer. When water can no longer travel through rock, it will travel sideways. In real life, this water will often move sideways underground until it reaches a stream, river, or lake.

How is groundwater used by people? **Wells** are dug into aquifers, and the water is pumped to the surface. This freshwater is used by homes, schools, and businesses. Farmers use groundwater to irrigate, or water, their crops.

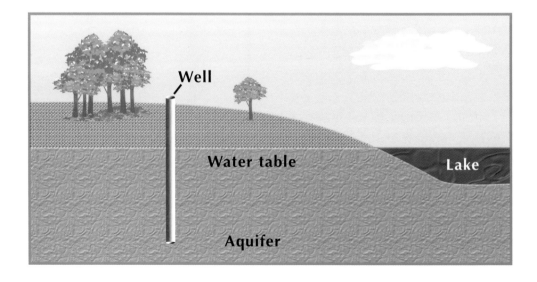

Surface Water

Lakes, rivers, and streams are examples of fresh surface water. Some of this water comes directly from precipitation. Rain, snow, or ice falls into a lake or river, adding to the water already there.

But surface water can also come from water that falls on the ground and makes its way to a lake, river, or stream. Precipitation that isn't absorbed by the ground becomes **runoff**. Runoff moves downhill and winds up in streams, rivers, lakes, and eventually the ocean.

Areas that have more pavement and less grass/soil have more runoff. Mountains and hills also have greater amounts of runoff. In addition to contributing to surface water sources, runoff also changes the face of the Earth as it wears down rock and soil and carries it to new locations.

Surface water is used much like groundwater. It is pumped or channeled to storage areas, where it is then **purified** for use by homes, businesses, and farms. Cities and towns near lakes and rivers often get much of their water from these sources.

8

The Cleaning Cycle

While all of the water on Earth is moving around, humans are constantly pulling water out of the cycle to use. But most of the time, the water they drink, cook with, bathe in, and water their gardens with isn't used straight out of a lake, river, or aquifer. Much of this water is contaminated with dirt, chemicals, or harmful organisms. It must be treated, or cleaned, before use. After use, it must be treated again before returning to the cycle. This treatment process is a small cycle within the larger water cycle.

Where Does Your Water Come From?

Water used by homes and businesses can come from one of several sources. It can come from a lake or river. It can come from a **reservoir**, or storage area for water. A reservoir may be a lake created by a dam or a big tank used to store large quantities of water. Water can also be pumped from underground wells. Water from wells may be pumped to storage tanks for cities or drawn up in smaller amounts for individual homes.

Reservoir in Utah

People who live in the country may have their own wells. They may add chemicals to their water to kill germs or use filters on their faucets to strain out harmful material.

People who live in a city are connected to the city's water supply. This water is treated in a water plant before being sent to homes and businesses. Water plants take water from its source and send it through a series of steps in order to purify it. There are many different ways to treat water. One common method is to mix chemicals with the water. These chemicals destroy any bad tastes or odors. Alum is one chemical often added to water during treatment. Alum causes "bad" particles to stick together and form globs called *floc*. These particles then settle to the bottom of the tank. The rest of the water is filtered through sand and gravel. Any remaining particles are strained out as the water passes through. Once the water is clean, chlorine is added to kill any bacteria that may be left. Many cities also add fluoride, which protects the teeth of the people who drink the water.

Now the water is ready to be distributed into the community. The clean water is sent to reservoirs or water towers. From there it travels through pipes to homes and businesses.

Water Treatment

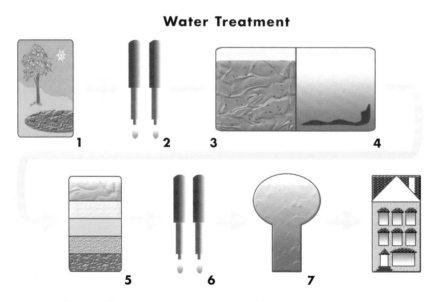

1. Water is drawn in from its source. 2. Alum is added to the water. 3. Floc forms. 4. Floc settles to the bottom of the tank. 5. Water is filtered through sand and gravel. 6. Chlorine and fluoride are added. 7. Water is stored until distributed to homes and businesses.

Towers of Water

How does a water tower work? Water is pumped out of the water treatment plant through pipes. Any water that isn't used immediately collects in the tower, which is just an open space in the main pipeline. Pipes run from the tower out to buildings in the community. When faucets in these buildings are turned on, they draw water out of the pipes. The water in the tower then replaces the water used from the faucet. Water is continually pumped into the tower from the treatment plant, so the flow of water isn't interrupted.

Where Does It Go When You're Done with It?

What happens to water after you've used it? You brush your teeth and take a shower. The dirty water goes down the drain. But where does it go from there? When your dishwasher or laundry machine empties dirty water into pipes, where does it go?

If you live in the country, your used water may travel through pipes into a septic tank that is buried underground. Solid waste sinks to the bottom of the tank. The water at the top flows out to a drainage field. When the waste compartment of the tank gets full, it must be emptied and cleaned. The waste material is then disposed of properly.

If you live in the city, your home is connected to the sewer system. Used water is piped out of buildings into bigger pipes under city streets. The large pipes lead to tanks at the city's **sewage** treatment plant. At the treatment plant, large objects are filtered out with screens. Smaller waste materials settle out and sink to the bottom of the tank. Air is then pumped into the wastewater so bacteria can do their work. These microscopic creatures break down remaining waste. The bacteria and wastes (sludge) then settle out. Chemicals such as chlorine are added to help kill any germs. The clean water is returned to a lake or stream. Treatment methods can vary from plant to plant, but the results will be the same.

> **Another Name for Dirty Water**
>
> Sewage plants are also called *wastewater plants*.

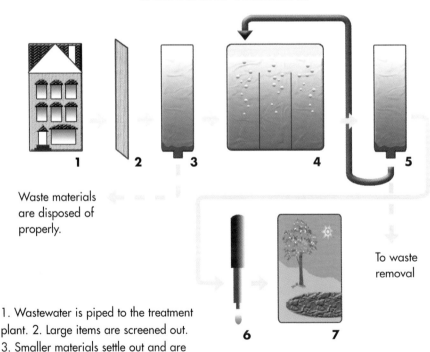

Wastewater Treatment

Waste materials are disposed of properly.

1. Wastewater is piped to the treatment plant. 2. Large items are screened out. 3. Smaller materials settle out and are removed. 4. Water is mixed with the oxygen so bacteria can grow and break down remaining waste. 5. Bacteria and waste settle out. Sludge is pumped back to the aeration tank or a waste tank. 6. Chlorine is added to kill bacteria. 7. Water is returned to the original source.

Thanks to the Cycle

Humans play a large role in the water cycle. They collect water, treat it, store it, and distribute it for use. Luckily, nature is able to recycle this water and put it back into the water cycle. If you do your part to conserve and protect water, the circle of water on Earth will provide you with many summer vacations filled with cool drinks, refreshing showers, and hours of fun at the pool.

Internet Connections and Related Reading for the Water Cycle

http://observe.arc.nasa.gov/nasa/earth/hydrocycle/hydro2.html
This NASA site takes you through the water cycle step by step. Moving diagrams demonstrate each stage. A water cycle quiz and word search puzzle show how much you've learned.

http://www.epa.gov/region07/kids/drnk_b.htm
Where does your drinking water come from? The Environmental Protection Agency answers this question for kids.

http://ga.water.usgs.gov/edu/mearth.html
Explore the Earth's water. Click on any of the variety of water topics including the water cycle, surface water, groundwater, and water distribution.

http://www.kidzone.ws/water/
Zone in on the water cycle with this simple overview and several printable activity sheets.

http://enchantedlearning.com/subject/astronomy/planets/earth/Watercycle.shtml
Check out the facts on the water cycle, and then complete a diagram, take a quiz, or put on a play.

A Drop of Water: A Book of Science and Wonder by Walter Wick. Describes the origins, characteristics, and uses of water. Scholastic, 1997. [RL 4 IL 4–8] (5755906 HB)

The Magic School Bus at the Waterworks by Joanna Cole. Eccentric Ms. Frizzle takes her class on a magical field trip to the waterworks in this fact-filled science adventure. Scholastic, 1986. [RL 3 IL 1–6] (8791301 PB 8791302 CC)

Water: Up, Down, and All Around by Natalie M. Rosinsky. Describes the water cycle and the importance of water, explaining evaporation and condensation, dew and frost, and the three states of water. Picture Window Books, 2003. [IL K–4] (3429106 HB)

- RL = Reading Level
- IL = Interest Level

Perfection Learning's catalog numbers are included for your ordering convenience. PB indicates paperback. CC indicates Cover Craft. HB indicates hardback.

Glossary

altitude (AL tuh tood) height above the surface of the Earth

aquifer (AH kwi fer) underground rock layer filled with water

cell (sel) smallest unit of life

cloud (klowd) mass of water or ice particles in the air

condensation (kon den SAY shuhn) process of water changing from a gas to a liquid; water droplets formed through condensation

convection current (kuhn VEK shuhn KER int) upward movement of warm air in the atmosphere

dilute (duh LOOT) to make something thinner or weaker by adding water

evaporation (ee vap uh RAH shuhn) process of water changing from a liquid to a gas

freshwater (FRESH waw ter) water containing very little or no salt

groundwater	(GROWND waw ter) water that soaks into the soil and collects underground
humid	(HYOU mid) having a lot of moisture in the air
ice cap	(ice kap) thick, permanent covering of ice and snow found at the North and South Poles and on mountains
molecule	(MAHL uh kyoul) smallest unit of a substance that can exist by itself
precipitation	(pree sip uh TAY shuhn) water that falls from the clouds in the form of rain, snow, freezing rain, sleet, or hail
purify	(PYOUR uh feye) to remove harmful substances
reservoir	(REZ uh vwar) area where water is stored
runoff	(RUHN awf) precipitation that flows across the ground into surface waters instead of soaking into the ground (see separate entries for *precipitation* and *surface water*)

salt water (sawlt WAW ter) water containing a significant amount of salt

sewage (SOO ij) waste material

surface water (SER fuhs WAW ter) water found on the surface of the Earth, such as oceans, lakes, and rivers

transpiration (trans puh RAY shuhn) evaporation of water from plants (see separate entry for *evaporation*)

water cycle (WAW ter SEYE kuhl) constant circulation of water from the air to the land to the sea; also called *hydrologic cycle*

water vapor (WAW ter VAY per) water in the form of a gas

well (wel) hole or passage that is dug or drilled into the ground in order to draw up water

Index

condensation, 24–29
    clouds, 24, 27–29
    fog, 25–26
contrails, 29
convection currents, 27
deposition, 17
desalination, 10
evaporation, 19–22
evapotranspiration, 23
freshwater, 8, 11
groundwater, 35–36
    aquifers, 35, 36
    water table, 35
precipitation, 30–33
    freezing rain, 32
    hail, 33
    rain, 32, 33
    sleet, 32
    snow, 31

reservoirs, 39
salt water, 8, 9–10
    Great Salt Lake, 9
sewage (wastewater) treatment, 41–42
states of water, 15–17
    gas, 16
    liquid, 16
    solid, 16
sublimation, 17
surface water, 37
    runoff, 37
transpiration, 22–23
water conservation, 12
water cycle
    definition of, 6, 7, 14–15
    diagram of, 18
water pollution, 13
water towers, 41
water treatment, 38–40